攀枝花市2013年度应用技术研究项目"国外钒钛研究重要文献摘要翻译及数据库建设"（编号：2013CY-R-13）阶段性研究成果

钒钛研究关键词中英文手册

主　编◎廖　红
副主编◎高朝阳　王海涛

西南交通大学出版社
·成都·

图书在版编目（CIP）数据

钒钛研究关键词中英文手册 / 廖红主编. —成都：西南交通大学出版社，2016.8
ISBN 978-7-5643-4663-8

Ⅰ. ①钒… Ⅱ. ①廖… Ⅲ. ①钒–稀有金属–金属材料–关键词–手册–汉、英②钛–轻有色金属–金属材料–关键词–手册–汉、英 Ⅳ. ①TG146-62

中国版本图书馆 CIP 数据核字（2016）第 085673 号

钒钛研究关键词中英文手册

主编 廖 红

责 任 编 辑	赵玉婷
封 面 设 计	墨创文化
出 版 发 行	西南交通大学出版社 （四川省成都市二环路北一段 111 号 西南交通大学创新大厦 21 楼）
发 行 部 电 话	028-87600564　028-87600533
邮 政 编 码	610031
网　　　　址	http://www.xnjdcbs.com
印　　　　刷	四川煤田地质制图印刷厂
成 品 尺 寸	148 mm×210 mm
印　　　　张	5.5
字　　　　数	152 千
版　　　　次	2016 年 8 月第 1 版
印　　　　次	2016 年 8 月第 1 次
书　　　　号	ISBN 978-7-5643-4663-8
定　　　　价	28.00 元

图书如有印装质量问题　本社负责退换
版权所有　盗版必究　举报电话：028-87600562

Preface 前言

  本手册编写的目的是为方便钒钛研究人员及相关专业人士查询中文关键词的英文表达。关键词主要涉及钒钛研究以及与钒钛材料应用紧密相关的领域，具体包括材料、冶金、化工、医药等。为了便于检索，统一按照中文关键词的拼音首字母顺序编排。

  编写时，以钒、钛、钒钛为关键词查询中国知网中涉及钒钛研究的期刊文章，将收集的文章中的关键词进行分类编排，同时对关键词的英文表达进行甄别。相同或相近的中文关键词有不止一种英文表达，在保证正确的前提下，对这些表达均予以保留。英文表达中的简写、符号也一并保留。

  本手册在编写过程中得到了攀枝花学院材料学院廖先杰博士的鼎力帮助，在此表示感谢。

  由于编者水平有限及编写时间仓促，书中难免存在错漏之处，敬请广大读者批评指正。

编 者

2015 年 10 月

Contents

A ······ 1
B ······ 2
C ······ 8
D ······ 18
E ······ 30
F ······ 32
G ······ 46
H ······ 61
J ······ 70
K ······ 82
L ······ 88
M ······ 97
N ······ 100
O ······ 105
P ······ 106
Q ······ 110
R ······ 114
S ······ 121

T 130
W 142
X 147
Y 153
Z 162

A

- 氨水沉钒
  vanadium precipitation with ammonia
- 螯合树脂 A
  chelate resin A
- 奥氏体
  austenite
- 澳矿
  Australian iron ore

钒钛研究关键词中英文手册

B

- 巴西铁精粉
  Brazil iron concentrates powder
- 靶
  target
- 靶衬间距
  substrate to target distance
- 白口铸铁
  white cast iron
- 白马钒钛磁铁矿
  Baima vanadium-bearing titanomagnetite
- 白马钒钛精矿
  Baima V-Ti concentrate
- 斑岩铜矿
  porphyry copper deposit
- 板坯
  slab
- 板钛矿
  brookite
- 半导体可饱和吸收镜
  semiconductor saturable absorber mirror
- 半密闭式钛渣电炉
  semi-hermetic titanium slag furnace
- 包覆
  coating

Bb

- 包利斯炉还原
  reducing with BORIS furnace
- 包装用镀铝薄膜
  plastic packaging film deposited with aluminium
- 保护覆盖层
  protective coating
- 保护气氛浇注
  pouring under controlled atmosphere
- 保护气氛热处理
  heat treatment in protective gas
- 爆炸焊
  explosion welding
- 饱和蒸汽压
  saturated vapor pressure
- 保护性
  protective
- 爆炸力学
  explosion mechanics
- 爆炸-轧制复合
  explosion and rolling cladding
- 贝氏体等温处理
  isothermal bainite treatment
- 贝氏体钢
  bainite steel
- 贝氏体等温淬火介质
  bainite austempering medium
- 贝氏体球墨铸球铁磨球
  bainite nodular cast iron grinding ball
- 焙烧
  roasting

钒钛研究关键词中英文手册

- 焙烧处理
  activated roasting
- 焙烧钒矿
  roasted vanadium ore
- 焙烧竖炉
  roasting shaft furnace
- 焙烧温度
  roasting temperature
- 本构模型
  constitutive model
- 便携X荧光光谱
  portable X-ray fluorescence spectrometer
- 变形
  deformation
- 变质处理
  modification treatment
- 变质剂
  modifier
- 变质作用
  metamorphism
- 标称转化效率
  specific conversion efficiency
- 标准曲线法
  standard curve method
- 表面处理
  surface treatment
- 表面氮化
  surface nitridation
- 表面等离子体
  surface plasma

- 表面电性
  surface potential
- 表面反应层
  surface reaction layer
- 表面复合层
  surface composite layer
- 表面改性
  surface modification
- 表面改性含钛炉渣
  surface modified titanium-bearing blast furnace slag
- 表面感应淬火
  surface induction hardening
- 表面化学处理
  surface chemical treatment
- 表面活性点
  surface-active site
- 表面活性剂
  surfactant
- 表面活性型助滤剂
  active surface filter aid
- 表面机械研磨(SMAT)
  surface mechanical attrition treatment (SMAT)
- 表面纳米化
  surface nano-crystallization
- 表面强化
  surface strengthening
- 表面润湿性
  surface wettability
- 表面渗钒
  superficial vanadizing

- 表面形貌
  surface morphology
- 表面性质
  surface properties
- 表面性状
  surface characteristics
- 表面修饰
  surface modification
- 表外矿
  ore blow cut-off grade
- 表征
  characterization
- 闭塞锻造
  core forging
- 变形铜合金
  deforming copper alloy
- 变形永磁钢
  deformable permanent-magnet steel
- 标准块
  reference block
- 表层纸状氧化物
  surface finger oxide
- 表面淬火
  surface hardening
- 表面改性材料
  surface modification material
- 表面热处理
  surface heat treatment
- 不规则状
  irregular

Bb

- 不完全退火
  partial anealing
- 不锈钢复合钢板和钢带
  stainless steel clad plate and strip
- 不锈钢丝
  stainless steel wire
- 布拉格定律
  the Bragg Law
- 部分合金化粉
  partially alloyed powder
- 波幅比
  wave amplitude ratio
- 波速
  wave velocity
- 玻璃转变温度
  glass transition temperature
- 薄膜
  thin film
- 不确定度
  uncertainty
- 不同类型萃取剂
  a variety of extractants
- 不锈钢
  stainless steel

C

- 材料表面与界面
  surface and interface in the materials
- 材料合成与加工工艺
  synthesizing and processing technics
- 材料失效与保护
  materials failure and protection
- 参数
  parameter
- 参数分析
  parameter analysis
- 残碳量
  carbon residue content
- 残余/逆转变奥氏体量
  residual/reverted austenite content
- 残余奥氏体
  retained austenite
- 残余应力
  residual stress
- 残留奥氏体
  retained austenite
- 层间腐蚀
  layer corrosion
- 超导体
  superconductivity

- 超高纯度不锈钢
  superhigh-pure stainless steel
- 超声波焊
  ultrasonic welding
- 超声雾化
  ultrasonic gas-atomizing
- 沉淀粉
  precipitated powder
- 沉淀物腐蚀
  deposit corrosion
- 淬火
  quench hardening
- 淬火冷却
  quenching
- 淬火冷却介质
  quenching medium
- 操作参数
  operating parameters
- 操作技术
  operating technology
- 操作制度
  operation institution
- 草酸盐沉淀法
  oxalate co-precipitation method
- 侧向长螺钉
  lateral long screw
- 测定
  determination
- 测定方法
  determination methods

钒钛研究关键词中英文手册

- 层状辉长岩体
  layered gabbro mass
- 差分电子密度
  electron density difference
- 差热分析
  differential thermal analysis（DTA）
- 差异机理分析
  mechanism analysis of beneficiability difference
- 柴油机缸套
  cylinder liners of diesel engine
- 掺铂
  platinum doped
- 掺钒
  vanadium doped
- 掺钛蓝宝石激光器
  sapphire laser
- 掺碳
  carbon doped
- 掺杂改性
  doping modification
- 产额
  yield
- 产业布局
  industrial layout
- 产业化
  industrialization
- 产业集聚
  industrial agglomeration
- 产质量
  yield and quality

Cc

- 常规破碎
  conventional crushing process
- 常压化学气相沉积法
  atmospheric pressure chemical vapor deposition
- 常压浸出
  leaching
- 超纯铁素体不锈钢
  ultrapure ferritic stainless steel
- 超弹性
  superelasticity
- 超富集
  hyper-accumulation
- 超高速撞击
  hyper-velocity impact
- 超高应变率
  super-high strain ratio
- 超固相线液相烧结
  supersolidous liquid phase sintering
- 超快激光
  ultrafast laser
- 超强激光
  ultraintense laser
- 超声波
  ultrasonic
- 超声化学
  sonochemistry
- 超细
  ultrafine
- 超细晶
  ultra-fine grains

钒钛研究关键词中英文手册

- 超细晶工业纯钛
  ultrafine-grained CP-Ti
- 超细晶粒钛
  ultrafine-grained Ti
- 超细晶组织
  ultra-fine microstructure
- 超细陶瓷色料
  ultrafine ceramic pigment
- 超重力场反应加工
  reaction processing in high-gravity field
- 车轴钢
  axle steel
- 沉淀法
  precipitation
- 沉淀强化
  precipitation hardening
- 沉淀相
  precipitation phase
- 沉积钒矿床
  sedimentary vanadium deposit
- 沉积物
  sediment
- 成骨细胞
  osteoblasts
- 成矿
  mineralization
- 成矿地质条件
  ore-forming geological conditions
- 成矿过程模拟
  ore-formation process modeling

Cc

- 成品率
  yield rate
- 成形能力
  shaping ability
- 成形性
  formability
- 程序升温表面反应
  temperature-programmed surface reaction
- 齿轮强度
  gear strength
- 赤铁矿-气孔结构
  hematite-pore structure
- 充放电控制
  charge-discharge control
- 充放电特性
  charge/discharge characteristics
- 充填材料
  filling materials
- 冲击磨损
  impact wear
- 冲击韧性
  impact toughness
- 臭氧
  ozone
- 出溶
  exsolution
- 出铁沟
  iron runner

▶13

- 初生相
  primary phase
- 初始组织
  initial structure
- 初渣
  initial slag
- 除硅
  de-silication
- 除磷
  de-phosphorization
- 除铁
  removing iron
- 储能
  energy storage
- 储氢合金材料
  hydrogen storage alloy
- 传感器
  sensor
- 传热分析
  heat transfer analysis
- 纯度
  purity
- 纯钙钛矿相
  pure perovskite phase
- 纯钛
  pure titanium
- 纯钛烤瓷冠
  porcelain fused to pure titanium

Cc

- 磁测资料
  magnetic data
- 磁电
  magnetoelectric
- 磁电耦合
  magnetoelectric coupling
- 磁电效应
  magnetoelectric effect
- 磁滑轮预选
  preliminary dressing by magnetic pulley
- 磁控溅射
  magnetron sputtering
- 磁铁矿
  vanadium titano-magnetite
- 磁性材料
  magnetic matetial
- 磁性特征
  magnetic property
- 磁选
  magnetic separation
- 磁选分离
  magnetic concentration
- 磁选柱
  column magnetic separator
- 磁学性质
  magnetic properties
- 粗糙度
  roughness

钒钛研究关键词中英文手册

- 粗粒抛尾
  tailing discarding at a coarser size
- 粗细分选
  discarding tailings at the coarser condition
- 催化
  catalysis
- 催化动力学光度法
  catalytic kinetic spectrophotometry
- 催化光度法
  catalytic spectrophotometry
- 催化还原
  catalyzing reduction
- 催化活性
  catalytic activity
- 催化剂
  catalyst
- 催化剂载体
  catalyst carrier
- 催化燃烧
  catalytic combustion
- 催化效率
  catalyst efficiency
- 催化荧光猝灭法
  catalytic fluorescence quenching method
- 淬透性
  hardenability
- 萃取
  extraction

Cc

- 萃取率
  extraction yield
- 萃原液
  extraction solution
- 措施
  measures

D

- 大规格棒材
  big size steel stick
- 大火成岩省
  large igneous provinces
- 大型钒钛磁铁矿床
  large vanadium titano-magnetite deposit
- 大型高炉
  large sized blast furnace
- 带有立筋板支撑炉腰托圈的炉体结构
  the body structure with vertical supporting plates upholding the furnace waist
- 单分子层
  monolayer spreading structure
- 单位质量热效应
  unit reaction heat
- 单向压缩
  unidirectional compression
- 单一磁选
  single magnetic separation
- 担载钒基氧化物
  supported-vanadium oxide
- 弹性性质
  elastic properties
- 蛋白

Dd

albumen
- 氮掺杂
  nitrogen-doped
- 氮等离子焰
  nitrogen plasma flame
- 氮氟掺杂
  nitrogen fluoride-codoped
- 氮化层
  nitrided layer
- 氮化钒
  vanadium nitride
- 氮化钒铁
  nitrided ferrovanadium
- 氮化钛
  titanium nitride
- 氮化钛薄膜
  titanium nitride film
- 氮流量
  nitrogen flow
- 氮收得率
  nitrogen yield
- 氮氧化物
  nitrogen oxides
- 导热系数
  thermal conductivity
- 导水裂隙带
  water flowing fractured zone
- 捣打料
  rammed mixture
- 等厚度图

isopach chart
- 等径弯曲通道变形
  equal channel angular pressing
- 等离子切割
  plasma cutting
- 等离子渗氮
  plasma nitriding
- 等离子渗镍
  plasma nickelizing
- 等离子束旋转电极
  plasma rotating electrode process
- 等离子体
  plasma
- 等体积浸渍法
  equal volume impregnation
- 打底焊
  backing welding
- 带状组织
  banded structure
- 等温锻
  isothermal forging
- 等温退火
  isothermal annealing
- 等温正火
  isothermal normalizing
- 低合金钢
  low-alloy steel
- 低强钛合金
  low-strength titanium alloy
- 低钛高碳锰铁

Dd

high carbon ferrochrome with low titanium
- 低碳钢
  low carbon steel
- 低温奥氏体不锈钢
  cryogenic austenitic stainless steel
- 低温回火
  low-temperature tempering
- 低温钛合金
  cryogenic titanium alloy
- 低温盐
  salt for low-temperature bath
- 电弧喷涂
  arc spraying
- 电化学腐蚀
  electrochemical corrosion
- 电解液淬火
  electrolytic hardening
- 定时淬火
  time quenching
- 定位焊
  tack welding
- 定影
  fixing
- 堆焊
  surfacing
- 低钒钢渣
  slag containing low-vanadium oxide
- 低铬白口铸铁
  low chromium white cast iron
- 低铬钒钛铸铁

low chromium vanadium titanium cast iron
- 低硅烧结
low silicon sintering
- 低耗
low consumption
- 低频冶炼
low frequency smelting
- 低品位
low grade
- 低品位钒钛磁铁矿
low-grade vanadium-titanium magnetite
- 低钛烧结矿
low-Ti sinter
- 低碳贝氏体钢
low carbon bainitic steel
- 低温
low temperature
- 低温焙烧
low-temperature roasting
- 低温催化剂
low-temperature catalyst
- 低温还原
low temperature reduction
- 低温还原法粉化
low temperature reduction powdering
- 低温还原粉化率
low temperature reduction degradation index
- 低温燃烧合成
low temperature combustion synthesize
- 低温制备

Dd

low temperature preparation
- 低氧高钛铁
  high titanium ferrous with low oxygen
- 低重复频率
  low repetition rate
- 低周疲劳
  low cycle fatigue
- 滴落带
  dropping zone
- 滴状冷凝
  dropwise condensation
- 锑
  antimony
- 地球化学样品
  geochemical samples
- 地球物理探矿
  geophysics prospecting
- 地质特征
  geological feature
- 地质样品
  geological samples
- 第一性原理
  first principles
- 点腐蚀性能
  pitting resistance
- 碘
  iodine
- 电场激活
  field activated
- 电池性能

▶23

battery performance
- 电催化
  electrocatalytic
- 电催化活性
  electrocatalysis properties
- 电催化氧化
  electrocatalytic oxidation
- 电导法
  electric conductivity method
- 电导率
  conductivity
- 电感耦合等离子体-原子发射光谱法
  Inductively Coupled Plasma-Atomic Emission Spectrometry (ICP-AES)
- 电感耦合等离子体质谱
  inductively coupled plasma mass spectrometry
- 电弧炉
  arc furnace
- 电弧烧损
  arc erosion
- 电化学性能
  electrochemical performance
- 电化学沉积
  electrochemical deposition
- 电化学活性
  electrochemical activity
- 电化学可逆性
  electrochemical reversibility
- 电化学性能
  electrochemical performance
- 电化学阻抗谱

Dd

electrochemical impedance spectroscopy
- 电火花沉积
electrospark deposition
- 电极材料
electrode materials
- 电极能耗
electrode consumption
- 电极失效
electrode degradation
- 电极寿命
electrode life
- 电解沉积
electrolytic deposition
- 电解还原
electrolytic reduction
- 电解液
electrolyte
- 电解液浓度
electrolyte composition
- 电解装置结构
electrolytic equipment structure
- 电绝缘技术
electrical insulation technology
- 电流密度
current density
- 电炉
electric furnace
- 电炉熔炼
electric furnace and smelting
- 电偶腐蚀

galvanic corrosion
- 电热熔分
electro-heat melting separation
- 电渗析
electro-dialysis
- 电输运性质
electrical transport
- 电脱氧
electro-deoxidation
- 电吸附
electro-sorption
- 电选
electro-static separation
- 电学性能
electrical properties
- 电压
voltage
- 电泳沉积
electrophoretic deposition
- 电子结构
electronic structure
- 电子理论
electronic theory
- 电子能带
electronic energy band
- 电子寿命
lifespan of electron
- 电子顺磁共振
electron paramagnetic resonance
- 电子态密度

Dd

density of electronic states
- 电子探针
  electron probe
- 电子显微分析
  electron microanalysis
- 电子自旋共振
  electron spin resonance
- 电阻率
  resistivity
- 调质工艺
  quenching and tempering process
- 动力学
  kinetics
- 动力学光度法
  kinetic spectrophotometry
- 动力学模型
  kinetics model
- 动力学特征
  dynamic characters
- 动态参数
  dynamic parameters
- 动态反应过程
  dynamic reaction process
- 动态吸附
  dynamic adsorption
- 动态析出
  dynamic precipitation
- 动态再结晶
  dynamic recrystallization
- 镀锌钢板

zinc-plated steel plate
- 短流程提钒
  short flow vanadium extraction
- 断裂
  fracture
- 断口分析
  fracture analysis
- 煅烧温度
  calcined temperature
- 锻造
  forging
- 对比试验
  comparison test
- 对策
  countermeasure
- 钝化
  passivation
- 多壁碳纳米管
  multi-walled carbon nanotubes
- 多尺度多层次复合
  multiscale and multilevel composite
- 多钒酸铵
  ammoninm poly-vanadate
- 多弧离子镀
  multiarc ion plating
- 多聚钒酸铵
  ammonium poly-vanadate
- 多孔 NiTi 形状记忆合金
  porous NiTi shape memory alloy
- 多孔材料

porous materials
- 多孔毛细芯
  porous wick
- 多孔纳米钛酸锶钡
  porous nano-barium-strontium titanate
- 多孔镍钛合金
  porous NiTi alloy
- 多孔钛
  porous titanium
- 多孔陶瓷膜
  porous ceramics film
- 多孔性
  porosity
- 多孔支架
  porous scaffold
- 多铁性材料
  multiferroics
- 多元非线性回归
  multiple regression
- 多元素钒钛球墨铸铁
  multi-alloys-containing V-Ti nodular

E

- 颚板
  jaw plates
- 二次磨矿
  secondary grinding
- 二次综合利用
  secondary comprehensive utilization
- 二茂钛二甲硫氨酸配合物
  titanocene di(L-methionine) dichloride
- 二茂钛水杨酸配合物
  dicyclopentadienyl salicylato-titanium(VI) complexes
- 二氧化钒
  vanadium dioxide
- 二氧化钒薄膜
  vanadium dioxide films
- 二氧化铅
  lead dioxide
- 二氧化钛
  titanium dioxide
- 二氧化钛薄膜
  TiO_2 films
- 二氧化钛纳米管
  TiO_2 nanotubes
- 二氧化钛纳米管阵列
  TiO_2 nanotube array

Ee

- 二氧化钛溶胶
  titania sol
- 二氧化钛陶瓷膜
  titanium dioxide ceramic film
- 二元碱度
  binary basicity
- 二次马氏体
  secondary martensite
- 二次硬化
  secondary hardening
- 二段正火
  two-step normalizing

F

- 钒
  vanadium
- 钒掺杂
  vanadium dope
- 钒萃取
  extraction vanadium
- 钒氮非调质钢
  V-N Non-Quenched and Tempered Steel
- 钒氮合金
  vanadium-nitrogen alloy
- 钒氮微合金钢
  vanadium-nitrogen microalloy steel
- 钒氮微合金化
  vanadium-nitrogen microalloyed
- 钒镀层
  vanadium coating
- 钒赋存状态
  occurrence of vanadium
- 钒铬物料
  vanadium-chromium materials
- 钒回收
  vanadium recovery
- 钒基催化剂
  vanadium-based catalysts

- 钒基固溶体

    vanadium-based (V-based) solid solution
- 钒基锐钛矿型催化剂

    vanadium-based (V-based) anatase catalysts
- 钒浸出率

    vanadium leaching rate
- 钒矿

    vanadium mine
- 钒矿石

    vanadium ore
- 钒离子

    vanadium ion
- 钒离子渗透

    vanadium ions permeation
- 钒离子渗透率

    vanadium ion permeability
- 钒离子指示电极

    vanadium ion indicator electrode
- 钒配合物

    vanadium complexes
- 钒迁移

    vanadium migration
- 钒取代

    vanadium-substitution
- 钒收率

    vanadium recovery
- 钒酸钙

    calcium vanadate
- 钒酸盐

    vanadate

钒钛研究关键词中英文手册

- 钒酸盐晶体
  vanadate crystals
- 钒酸盐荧光粉
  vanadate phosphor
- 钒酸钇晶体
  YVO_4 crystal
- 钒钛
  V-Ti
- 钒钛磁铁精矿
  vanadium-bearing titanomagnetite concentrates
- 钒钛磁铁精矿烧结
  vanadium titanium magnetite concentrate sintering
- 钒钛磁铁矿煤球团
  Vanadic-titanomagnetite-coal mixed pellet
- 钒钛磁铁矿球团
  V-Ti magnetite pellet
- 钒钛磁铁铁矿床
  V-Ti magnetite deposit
- 钒钛催化剂
  TiO_2 catalysts
- 钒钛低合金抗磨白口铸铁
  wear-resist white cast iron of low vanadium-titanium alloy
- 钒钛粉
  vanadium-titanium powder
- 钒钛高炉渣
  blast furnace slag with vanadium-titanium
- 钒钛黑瓷
  V-Ti black ceramics
- 钒钛回收
  recovery of vanadium and titanium

Ff

- 钒钛基地
  vanadium-titanium base
- 钒钛精矿
  vandium-bering titaniferous magnetite concentrate
- 钒钛矿
  vanadium titanium iron ore
- 钒钛矿及渣
  sefstromite or slags
- 钒钛矿冶炼
  smelting V-Ti bearing magnetite
- 钒钛连测
  vanadium-titanium continuous measurement
- 钒钛磷铸铁气缸套
  vanadium- titanium-phosphorus (V-Ti-P) cylinder liner
- 钒钛硫复合催化剂
  sulfur modified vanadia-titania-sulfur catalyst
- 钒钛炉渣
  vanadium- titanium (V-Ti) slag
- 钒钛球墨铸铁
  vanadium titanium spheroidai graphite cast iron
- 钒钛球团
  vanadium-titanium (V-Ti) pellets
- 钒钛蠕墨铸铁
  vanadium-titanium vermicular cast iron
- 钒钛烧结矿
  vanadium-titanium (V-T-i) bearing sinter
- 钒钛生铁
  vanadiumt itanium cast iron
- 钒钛生铁控制样品
  controling sample of vanadium titanium pig iron

▶35

钒钛研究关键词中英文手册

- 钒钛酸
  vanadium-titanium acid
- 钒钛酸性渣
  acidic vanadium-titanium slag
- 钒-钛陶瓷
  V_2O-TiO_2 ceramic
- 钒钛铁精矿
  vanadium and titanium iron concentrate
- 钒钛铁矿
  vanadium-titanium iron
- 钒钛铁水
  hot metal containing V and Ti(vanadium and titanium)
- 钒钛微合金钢
  T-Vi(vanadium and titanium) Microalloyed Steel
- 钒钛微合金化
  vanadium-titanium microalloy
- 钒钛物料护炉
  hearth protection with V-Ti bearing materials
- 钒钛氧化
  vanadium-titanium oxidation
- 钒钛冶炼
  vanadium titanium smelting
- 钒钛资源
  vanadium-titanium resources
- 钒铁掺杂
  vanadium and ferrum (V and Fe) co-doped
- 钒铁尖晶石
  vanadium iron spinel
- 钒微合金 TRIP 钢
  vanadium-containing TRIP steel

- 钒微合金化 TRIP 钢
  V-microalloyed TRIP aided steel
- 钒微合金化
  V microalloying
- 钒系催化剂
  vanadium catalyst
- 钒系磷酸盐
  vanadium-based phosphate
- 钒系氧化物
  vanadium oxides
- 钒氧化
  vanadium oxidation
- 钒氧化还原液流电池
  vanadium redox flow battery
- 钒氧化钠化焙烧
  vanadium oxidation and sodium change roasting
- 钒氧化物
  vanadium oxide
- 钒氧配合物
  oxovanadium complex
- 钒氧物种
  vanadium oxide
- 钒冶金废水
  vanadium metallurgical waste water
- 钒液流电池
  vanadium redox battery
- 钒元素
  vanadium
- 钒渣
  vanadium slag

- 钒渣直接合金化
  directly alloying with V and Ti slag
- 钒资源
  titanium resources
- 钒合金
  vanadium alloy
- 钒铝合金
  vanadium aluminum alloy
- 钒铝中间合金
  vanadium-aluminum master alloy
- 钒铁
  ferrovanadium
- 反极图
  inverse pole figure
- 反射率
  reflectivity
- 反应烧结
  reaction sintering
- 防腐剂
  preservation agent
- 非合金钢
  unalloyed steel
- 非合金工具箱
  nonalloy tool steel
- 非合金铸钢
  nonalloy cast steel
- 分层
  lamination
- 分级
  classification

- 分子光谱
  molecular spectroscopy
- 粉浆
  slurry
- 粉块
  cake
- 粉末锻造
  powder forging
- 粉末衍射卡
  powder diffraction file
- 粉末轧制
  power rolling
- 服役能力
  serviceability
- 服役寿命
  service life
- 腐蚀保护
  corrosion protection
- 腐蚀产物
  corrosion product
- 复合热处理
  duplex heat treatment
- 复合压坯
  composite compact
- 覆盖剂
  coverture
- 覆膜铁
  laminated steel
- 反萃
  back extraction

钒钛研究关键词中英文手册

- 反萃取
  back-extraction
- 反浮选
  reverse flotation
- 反钙钛矿结构
  anti-perovskite structure
- 反应机理
  reaction mechanism
- 反应离子刻蚀
  reactive ion etching
- 反应途径和步骤
  reaction route and step
- 方案
  scheme
- 仿生矿化
  biomimetic mineralization
- 仿生涂层
  biomimetic coating
- 仿真
  simulation
- 放电等离子烧结
  spark plasma sintering
- 飞秒激光
  femtosecond laser
- 非等温还原
  nonisothermal reduction
- 非钒基催化剂
  vanadium-free base catalysts
- 非高炉炼铁
  non-blast furnace ironmaking

Ff

- 非晶相
  amorphous phase
- 非均相芬顿反应
  non-homogeneous Fenton reaction
- 非均匀磁系
  non-uniform magnetic system
- 非水体系
  nonaqueous system
- 非线性弹性
  non-linear elastic deformation
- 废催化剂
  dead catalyst
- 废金属钛合金
  scrap titanium alloy
- 废水
  waste water
- 废水处理
  waste water treatment
- 废水蒸发
  waste water evaporation
- 废酸
  waste acid
- 废盐酸
  waste hydrochloric acid
- 分布
  distribution
- 分布板
  distribution plate
- 分离
  separation

钒钛研究关键词中英文手册

- 分离方法
  analytical methods
- 分流制粒
  separated granulating
- 分配行为
  distribution behaviour
- 分散
  dispersion
- 分散性
  dispersibility
- 分散液液微萃取
  dispersive liquid-liquid microextraction
- 分束棱镜
  splitting prisms
- 分选指标
  sorting index
- 分子动力学
  molecular dynamics
- 分子间氢键
  intermolecular hydrogen bonds
- 粉化
  efflorescence
- 粉末冶金
  powder metallurgy
- 粉末冶金工具钢
  PM (power metallurgy) tool steels
- 粉体
  powder
- 粉体特性
  power property

Ff

- 风化矿
  weathered ores
- 氟化镧
  lanthanum fluoride
- 氟钛酸钾
  potassium pluotitanate
- 浮选
  flotation
- 浮选分离
  flotation separation
- 辅料
  auxiliary material
- 腐蚀电流密度
  corrosion current density
- 腐蚀磨损
  corrosive wear
- 腐蚀深度
  corrosion depth
- 腐蚀形貌
  corrosion morphology
- 腐蚀性能
  corrosion
- 负极材料
  cathode material
- 负载型催化剂
  support atalysts
- 复合板
  clad plates
- 复合变质
  compound inoculates

钒钛研究关键词中英文手册

- 复合储氢材料
  hydrogen storage composite
- 复合锤头
  compound hammerhead
- 复合镀层
  composite coating
- 复合负载
  composite supported
- 复合-扩散连接
  composite-diffusion bonding
- 复合模板
  composite templates
- 复合膜
  composite coating
- 复合球团
  composite pellets
- 复合钛基润滑脂
  titanium complex grease
- 复合添加剂
  complex additives
- 复合填料
  composite filler
- 复合涂层
  composite coating
- 复合砖衬
  line of compound brick
- 复杂地层
  complex formation
- 富钒石煤
  vanadium-enriched stone coal

Ff

- 富集
  concentrate
- 富矿
  rich ore fines
- 富钛料
  rich titanium material
- 富钒渣
  vanadium-rich slag
- 赋存状态
  existence form
- 覆层
  cladding

G

- 钆
  gadolinium
- 改进的霍洛蒙模型
  revised Hollomon model
- 改良模拟体液
  modified simulated body fluid
- 改性
  modification
- 改性电极
  modified electrode
- 改性沸石
  modify zeolite
- 钙
  Ca(calcium)
- 钙比
  calcium ratio
- 钙化焙烧
  calcination with calcium compound
- 钙化沉钒
  calcification
- 钙热还原
  calciothermic reduction
- 钙钛锆石
  zirconolite

Gg

- 钙钛矿
  perovskite
- 钙钛矿催化剂
  perovskite catalyst
- 钙钛矿结构
  perovskite structure
- 钙钛矿氧化物催化剂
  perovskite-type oxide catalyst
- 钙钛矿氧化物异质结
  perovskite oxide heterojuction
- 干扰曲线法
  interfering curve method
- 甘氨酸–硝酸盐法
  glycine-nitrate process
- 刚玉
  corundum
- 钢
  steel
- 钢包
  ladle
- 钢锭模
  ingot mould
- 钢绞线
  steel strand
- 钢冷却壁
  steel cooling wall
- 钢铁样品
  iron and steel samples
- 钢枕模
  steel ingot mold

- 高氧化铝钒钛炉渣
  vanadium-titanium BF slag of high Al_2O_3 content
- 高钛铝合金
  high TiAl alloy
- 高产
  high yeild
- 高纯
  high-purity
- 高磁场强度
  high magnetic field strength
- 高碘酸钾
  potassium periodate
- 高钒高速钢
  high vanadium high speed steel
- 高钒铁
  high vanadium ferrovanadium
- 高负压
  high negative pressure
- 高富氧喷吹
  PC injection with high oxygen enrichment
- 高钙镁钛渣
  titanium slag high in calcium and magnesium
- 高铬
  high chromium content
- 高铬型钒钛磁铁矿
  high chromium vanadium-titanium magnetite
- 高铬铸铁
  high-chromium cast iron
- 高功率脉冲磁控溅射
  high power pulsed magnetron sputtering (HPPMS)

- 高硅高碳钒矿

  vanadium ore with high carbon and silicon content
- 高回收率

  high recovery
- 高级氧化

  advanced oxidation
- 高碱度

  high basicity
- 高阶色散

  high-order dispersion
- 高结晶度

  high crystallinity
- 高粱秸秆

  sorghum straw
- 高硫酸用量

  high dose of sulfuric acid
- 高炉长寿

  prolonging campaign
- 高炉热平衡

  blast furnace (BF) heat balance
- 高炉冶炼

  blast-furnace smelting
- 高炉渣

  blast furnace slag
- 高铝烧结矿

  high-alumina sinter
- 高铝中钛渣

  blast furnace slag with medium titanium and high alumina content
- 高锰钢

  high-manganese steel

- 高能反应球磨
  high energy reaction ball milling
- 高能晶面
  high energy facets
- 高能喷丸
  high-energy shot peening (HESP)
- 高能振动球磨
  high-energy vibratory ball milling
- 高频感应
  high-frequency induction
- 高品位钒钛磁铁精矿
  high grade vanadium-bearing titaniferous magnetite concentrate
- 高强钢筋
  high-strength reforcement bar
- 高强混凝土
  high strength concrete
- 高钛钢
  high titanium steel
- 高钛高炉渣
  high titanium-bearing slag
- 高钛铝土矿
  high titanium bauxite
- 高钛铁
  high titanium ferroalloy
- 高钛型钒钛矿冶炼
  high titanium-bearing ore
- 高钛型球团矿
  high titania type pellet
- 高钛渣
  titanium slag

Gg

- 高钛重渣
  high titanium heavy slag
- 高碳钢
  high carbon steel
- 高碳钢盘条
  high carbon wire and rod
- 高碳石煤
  high carbon stone coal
- 高铁低硅
  high-ferrum and low-silicon
- 高温
  high temperature
- 高温固相法
  solid-state reaction method
- 高温拉伸
  high temperature tensile
- 高温力学性能
  mechanical properties at elevated temperature
- 高温磨损
  high-temperature wear
- 高温屈服强度
  high temperature yield strength
- 高温熔化性质
  melting property in high temperature zone
- 高温试验
  high-temperature tests
- 高温塑性
  hot plasticity
- 高温钛合金
  high temperature titanium alloy

- 高温碳化
  high temperature carbonization
- 高温氧化
  high temperature oxidation
- 高效生产
  high productivity
- 高压
  high pressure
- 高压辊磨
  high pressure rolling process
- 高压辊磨机
  high pressure griding roller
- 高盐废水
  high-salt wastewater
- 高电阻电热合金
  high resistance alloy for electrical heating
- 高分辨电子显微术
  high-resolution electron microscopy
- 高炉锰铁
  blast furnace ferromanganese
- 高强度不锈钢
  high-strength stainless steel
- 高强度钢丝
  higher strength steel wire
- 高强钛合金
  high-strength titanium alloy
- 高速锻造
  high-velocity forging process
- 高钛钾型焊条
  high titanium potassium electrode

Gg

- 高钛钠型焊条
  high titanium sodium electrode
- 高温合金
  heat-resisting superalloy
- 高温回火
  high-temperature tempering
- 高温金相
  high-temperature metallography
- 高温盐
  salt for high-temperature bath
- 高温盐浴矫正剂
  high-temperature salt bath rectifier
- 高压电子显微术
  high-voltage electron microscopy
- 锆钛砂矿
  zirconium and titanium placer
- 锆掺杂
  zirconium doping
- 锆钛酸铅镧
  lead lanthanum zirconate titanate
- 镉
  cadmium
- 铬
  chromium
- 铬合金
  chromium alloy
- 铬镍共渗
  plasma nickel-chromium metalizing
- 铬酸钾
  potassium chromate (K_2CrO_4)

- 根管长度
  root canal length
- 根管润滑剂
  root canal lubricant
- 根管预备
  root canal preparation
- 根管再治疗
  endodontic retreatment
- 根管治疗
  root canal therapy
- 根际环境
  rhizosphere
- 根尖裂纹
  apical root cracks
- 工程特性
  engineering characteristic
- 工程应变
  engineering strain
- 工程应力
  engineering stress
- 工业纯铁
  ingot iron
- 工业二氧化钛
  nitrogen doped TiO_2
- 工业纯钛
  pure titanium
- 工业废弃物
  industrial waste
- 工业利用
  industrial utilization

Gg

- 工业试验
  industrial experiment
- 工业钛液
  industrial titanyl sulfate
- 工业指标
  industrial index
- 工艺
  technology
- 工艺参数
  technological parameters
- 工艺改进
  technological improvement
- 工艺矿物学
  mineralogy
- 工艺设计
  process design
- 工艺条件
  process condition
- 工艺性能
  processing property
- 工艺优化
  technology modification
- 工艺指标
  technique specifications
- 工艺制度
  process system
- 公司治理
  corporate governance
- 功率波动
  power fluctuation

- 功能梯度材料
  functionally graded materials
- 供氧量
  oxygen supply
- 汞
  mercury
- 共掺杂
  codoping
- 共存理论
  coexistence theory
- 骨结合
  osseointegration
- 钴
  cobalt
- 钴酸钡钐
  samarium barium cobalt oxide cathode
- 鼓风动能
  blast kinetic energy
- 固结机理
  bonding mechanism
- 固溶量
  solid-soluted content
- 固溶体
  solid solution
- 固态反应
  solid-state reaction
- 固体废物
  solid wastes
- 固体渗硼
  solid-state pack boronizing

Gg

- 固体透氧膜
  solid oxygen membrane (SOM)
- 固体氧化物燃料电池
  solid oxide fuel cells
- 固相烧结
  solid phase sintering
- 固相烧结法
  solid sintering technology
- 管材
  tube
- 管径
  nanotube diameter
- 管式炉
  tubular-furnace
- 管线钢
  pipeline steel
- 光催化
  photocatalytic
- 光催化还原
  photocatalytic reduction
- 光催化活性
  photocatalytic activity
- 光催化剂
  photocatalyst
- 光催化降解
  photocatalytic degradation
- 光催化协同作用
  photocatalytic synergistic effect
- 光催化性能
  photocatalytic performances

钒钛研究关键词中英文手册

- 光催化氧化
  photocatalytic oxidation
- 光电效应
  photoelectric effect
- 光电性质
  photoelectric properties
- 光度法
  spectrophotometry
- 光滑粒子法
  smoothed particle hydrodynamics method
- 光谱拟合
  spectrum fitting
- 光吸收
  optical absorption
- 光纤激光-MIG复合焊
  fiber laser-MIG hybrid welding
- 光纤激光焊
  fiber laser welding
- 光学常数
  optical constant
- 光学性能
  optical properties
- 光学性质
  optical property
- 光与物质相互作用
  interaction of light with matter
- 光亮热处理
  bright heat treatment
- 光亮退火
  bright annealing

Gg

- 广角
  wide angle
- 规划设计
  plan and design
- 硅
  silicon
- 硅钢
  silicon steel
- 硅含量
  Si content
- 硅酸盐相
  silicate phase
- 硅钛柱撑
  silicon-titanium pillared
- 硅烷化
  silanization
- 硅烷膜
  silane film
- 硅烷偶联剂 KH-550
  silane coupling agent KH-550
- 贵金属
  noble-metal
- 贵金属雾化粉
  atomized precious metal powder
- 国家矿山公园
  national mining park
- 过程控制
  process control
- 过渡金属
  transition metal

钒钛研究关键词中英文手册

- 过渡类型
  transitional type
- 过还原
  over reduction
- 过滤
  filtration
- 过氧钛系
  peroxy-titanium complex

H

- 海滨砂矿
  beach placer
- 海绵钛
  titanium sponge
- 海洋沉积物
  marine sediment
- 海藻酸盐
  alginate
- 氦-空位复合物
  helium-vacancy complex
- 含氮铁合金
  ferroalloy with nitrogen
- 含钒 TRIP 钢
  vanadium-bearing TRIP steel
- 含钒磁铁矿
  vanadium-bearing magnetite
- 含钒钢渣
  vanadium-bearing steel slag
- 含钒灰渣
  vanadium containing ash
- 含钒浸出液
  vanadium leaching solution
- 含钒矿物
  vanadium-bearing minerals

- 含钒钛 TRIP 钢
  TRIP steel containing vanadium and titanium
- 含钒钛钢渣
  slag with vanadium and titanium oxides
- 含钒铁水
  vanadium-bearing molten iron
- 含钒尾渣
  vanadium-containing residue
- 含钒页岩
  vanadium-containing shale
- 含钒粘土矿
  vanadium-bearing clay
- 含氟助浸剂
  fluorine-containing leaching agent
- 含钛复合矿
  titanium-bearing compounds
- 含钛高炉渣
  titanium-bearing blast furnace slag
- 含钛夹杂物
  inclusions containing titanium
- 含钛镍铬合金
  nickel-chromium-titanium porcelain alloy
- 含钛烧结矿
  titanium containing sinter
- 含钛铁基粉末
  titanium-containing iron-based powder
- 含钛铁素体不锈钢
  titanium-containing ferritic stainless steel
- 含碳球团
  carbon composite pellet

Hh

- 焊接
  welding
- 焊接接头
  welded joint
- 焊丝钢
  welding steel
- 焊条
  electrode
- 氧还原
  aerobic reduction
- 合成
  synthesis
- 合成过程
  synthesizing process
- 合金化
  alloying
- 褐铁矿
  limonite
- 黑色金属
  ferrous metal
- 黑色岩系
  black rock formation
- 黑钛石
  anosovite
- 痕量分析法
  trace analysis
- 痕量元素
  trace elements
- 恒温氧化动力学曲线
  isothermal oxidation kinetic curves

钒钛研究关键词中英文手册

- 横截面
  cross-section
- 烘干温度
  drying temperature
- 红景天
  rhodiola
- 红外光谱
  infrared spectrometry
- 红外探测器
  infrared detectors
- 红外透过率
  infrared transmittance
- 宏观硬度
  macro hardness
- 后处理
  after treatment
- 护壁堵漏
  wall protection and leakage control
- 护炉
  hearth lining protection
- 滑动面
  sliding surface
- 滑膜
  synovium
- 化铁炉渣
  cupola slag
- 还原渣
  reducing slag
- 化学成分
  chemical compositions

Hh

- 电化学传感器
  electrochemical sensor
- 化学镀
  chemical plating
- 化学改进剂
  chemical modifier
- 化学相分析
  chemical phrase analysis
- 化学形态
  chemical speciation
- 化学转化膜
  chemical conversion coating
- 化学状态
  chemical states
- 划痕实验
  scratch test
- 矿石还原
  ore reduction
- 还原产物
  reduction product
- 还原度
  degree of reduction
- 还原反应
  reduction reactions
- 还原机理
  reduction mechanism
- 还原历程
  reduction progress

- 还原膨胀
  reduction swelling
- 还原膨胀指数
  reduction swelling index
- 还原气氛
  reducing atmosphere
- 还原速度
  reduction rate
- 还原铁粉
  reduced iron powder
- 还原效应
  reduction effect
- 还原锈蚀法
  reduction-rust process
- 环己烷
  cyclohexane
- 环己烯环氧化
  epoxidation of cyclohexene
- 环境
  environment
- 环境工程
  environment engineering
- 环境矿物学
  environmental mineralogy
- 环境气体
  ambient gas
- 环境扫描电镜
  environmental scanning electron microscopy (ESEM)

Hh

- 环境效应
  environmental effect
- 环境指示意义
  environmental marker function
- 环境治理
  environment control
- 环境质量评价
  environmental quality assessment
- 环路热管
  loop heat pipe
- 环氧化
  epoxidation
- 缓冲矿仓
  product storage bin
- 灰铁
  gary iron
- 灰铸铁
  gary pig iron
- 挥发成分
  volatiles
- 辉长岩
  gabbro rock
- 回归正交设计
  orthogonal regression
- 回火工艺
  tempering process
- 回火温度
  tempering temperature

- 回收
  recovery
- 回收利用
  recycling
- 回转窑
  rotary kiln
- 回转窑结圈
  rotary kiln ringing
- 混合料水分
  moisture in mixture
- 混凝
  coagulation
- 混凝土
  concrete
- 混烧
  mixed combustion
- 活度
  activity
- 活度系数
  activity coefficient
- 活化焙烧
  activated roasting
- 活化烧结
  activate sintering
- 活化能
  activation energy
- 活塞环
  piston ring

- 活性石灰
  active lime
- 活性氧
  reactive oxygen species（ROS）
- 活性组分
  active loading
- 火法冶炼
  pyrogenic process
- 火焰原子吸收光谱法
  flame atomic absorption spectrometry (FAAS)

J

- 机理
  mechanism
- 机械化学
  mechanochemistry
- 机械性能
  mechanical properties
- 机用镍钛锉
  machine nickel-titanium file
- 机用镍钛器械
  nickel-titanium rotary instruments
- 积累
  accumulation
- 基础特性
  fundamental characteristics
- 基础性能
  basic property
- 基体
  matrix
- 激光沉积成形
  laser metal deposition
- 激光冲击形变
  laser shock deformation
- 激光淬火
  laser hardening

Jj

- 激光合金化
  laser alloying
- 激光等离子体
  laser plasma
- 激光工艺参数
  laser parameters
- 激光焊接
  laser welding
- 激光技术
  laser technique
- 激光晶体
  laser crystal
- 激光器
  lasers device
- 激光强化
  laser strengthening
- 激光熔覆
  laser cladding
- 激光熔化沉积
  laser melting deposition
- 激光透射连接
  laser transmission joining
- 激光诱导等离子体
  laser-induced plasma
- 激光直接融覆
  laser direct depositing
- 激活能
  activation energy
- 极差分析
  range analysis

- 极化电阻
  polarization resistance
- 极化强度
  intensity of polarization
- 极化曲线
  polarization curve
- 极贫钒钛磁铁矿
  lower grade vanadium-bearing titanomagnetite
- 极谱法
  polargraphy
- 急冷烧结法
  quench sintering process
- 集流体
  current collector
- 集群滑移
  stratified cluster slipping
- 挤压
  extrusion
- 计算法
  caculation method
- 计算机辅助设计
  computer aided design(CAD)
- 技术创新
  technology innovation
- 技术选择
  technological strategy choice
- 加工图
  processing map
- 加碳氯化
  carbon-chlorination

Jj

- 加压高温碱浸
  pressurized high temperature alkali leaching
- 夹砂
  scab
- 夹点技术
  pinch technology
- 夹杂物
  inclusion
- 镓
  gallium
- 甲醇选择氧化
  methanol selective oxidation
- 甲缩醛
  methylal
- 钾修饰
  K(kalium)-modification
- 尖晶石
  spinel
- 间隙氮
  interstitial nitrogen (N)
- 剪切强度
  shear strength
- 检测
  detect potential
- 碱比
  soda ratio
- 碱度
  basicity
- 碱金属
  alkali metal

钒钛研究关键词中英文手册

- 碱浸
  alkaline leaching
- 碱侵蚀
  alkali attack
- 碱热处理
  alkali-thermal-treatment
- 碱性辉长岩
  alkali gabbro
- 碱性磷酸酶
  alkaline phosphatase
- 建筑结构钢
  structural steel
- 溅射功率
  sputtering power
- 溅射气压
  sputtering pressure
- 溅射时间
  sputtering time
- 溅射压强
  sputtering pressure
- 溅渣护炉
  slag splashing
- 降解
  degradation
- 降解率
  degradation rate
- 交流阻抗
  alternating current impedance
- 胶束催化
  micellar catalysis

Jj

- 焦比
  coke ratio
- 焦粉用量
  coke breeze proportion
- 焦炭质量
  coke quality
- 角部横裂纹
  corner transversal crack
- 搅拌摩擦加工
  friction stir processing
- 阶段磨矿
  stage grinding
- 阶段磨矿分级
  stage grinding classification
- 阶段磨矿阶段选别
  stage-grinding and stage-separation
- 阶段磨选
  stage grinding and concentration
- 阶段弱磁
  stage low intensity magnetic separation
- 接触磨损
  contact wear
- 接头性能
  mechanical property
- 节能
  energy saving
- 节能方向和途径
  directions and measures of energy conservation
- 节能减排
  energy saving and emission reduction

钒钛研究关键词中英文手册

- 结构
  structure
- 结果不稳定
  unstable result
- 结合界面
  bonding interface
- 结合强度
  bonding strength
- 结晶
  crystallization
- 结晶分离
  crystallization separation
- 结晶质量
  crystal quality
- 解决措施
  solving measure
- 解离常数
  dissociation constant
- 解吸
  desorption
- 介电常数
  dielectric constant
- 介电损耗
  dielectric loss
- 介孔
  mesoporous
- 介孔材料
  mesoprous materials
- 介孔二氧化钛
  mesoporous titania

Jj

- 介孔分子筛
  mesoporous molecular sieve
- 介孔结构
  mesoporous structure
- 介质淬火
  medium quenching
- 界面波
  interface wave
- 界面反应
  interfacial reaction
- 界面极化电阻
  interfacial polarization resistance
- 界面乳化物
  interfacial emulsion
- 金红石
  rutile
- 金红石型钛白粉
  rutile titanium dioxide
- 金相组织
  metallographic structure
- 金相腐蚀
  corrosion metallography
- 金属材料
  metallic materials
- 金属掺杂
  metal doping
- 金属化
  metallization
- 金属化还原
  metallization reduction

钒钛研究关键词中英文手册

- 金属化率
  metallization degree
- 金属回收率
  recovery ratio of metal
- 金属基复合材料
  metallic matrix composites
- 金属间化合物
  intermetallic compound
- 金属钛夹
  titanium clips
- 金属铁
  metallic iron
- 金属氧化物
  metal oxides
- 金属氧化物涂层
  metal oxide coating
- 浸出
  leaching
- 浸出动力学
  leaching kinetics
- 浸出过程
  leaching process
- 浸出率
  leaching rate
- 浸钒
  leaching vanadium
- 浸硅
  leaching silicium
- 浸入式水口
  submerged entry nozzle

- 经济品位
  economic grade
- 经济效益
  economic benefit
- 晶格能
  lattice energy
- 晶界
  crystal boundary
- 晶界磁性势垒
  magnetic Schottky barrier
- 晶界取向差
  boundaries with misorientation
- 晶粒生长
  grain growth
- 晶粒尺寸
  grain size
- 晶粒粗化温度
  grain coarsening temperature
- 晶粒细化
  grain refinement
- 晶粒细化机制
  grain refinement mechanism
- 晶体结构
  crystal structure
- 晶体塑性
  crystal plasticity
- 晶型
  crystal type
- 晶型转变
  phase transition

钒钛研究关键词中英文手册

- 精矿品位
  concentrate grade
- 精炼剂
  refine agent
- 精炼渣
  refining slag
- 精馏
  distillation
- 净化
  purification
- 静电纺丝
  electrospinning
- 静态吸附
  static adsorption
- 静态再结晶
  static recrystallization
- 居里温度
  Curie temperature
- 局域表面等离子体共振
  localized surface plasmon resonance
- 剧烈塑性变形
  severe plastic deformation(SPD)
- 聚苯胺
  polyaniline
- 聚吡唑硼酸盐
  poly pyrazolyl borate
- 聚硅氮烷先驱体
  polysilazane precursor
- 聚硅硫酸钛
  poly-titanium-silicate-sulphate(PTSiS)

- 聚合硅酸钛铝
  poly-titanium-aluminum-silicate(PTAS)
- 聚合硫酸铁
  polymeric ferric sulfate
- 聚合钛离子
  poly-titanium-ion
- 聚偏离氟乙烯
  polyvinylidene fluoride (PVDF)
- 卷取温度
  coiling temperature
- 绝对冶炼时间
  absolute smelting time
- 绝热温度
  adiabatic temperature
- 均四甲苯
  durene
- 均匀化方法
  homogenization method
- 均匀设计
  uniform design

K

- 开发现状
  exploration situation
- 开裂
  crack
- 抗腐蚀磨损
  corrosive-wearing resistance
- 抗高温氧化性能
  high temperature oxidation resistance
- 抗菌性能
  antibacterial performance
- 抗拉强度
  tensile strength
- 抗磨钢
  abrasive resistance steel
- 抗汽蚀
  anti-cavitation
- 抗压强度
  compressive strength
- 抗氧化性
  oxidation resistance
- 抗震
  earthquake resistance
- 抗震设计
  anti-seismic design

Kk

- 抗震性能
  seismic performance
- 抗肿瘤活性
  antitumor activity
- 科技成果转化
  transformation of scientific and technological achievements
- 颗粒粒径
  particle size
- 可持续发展
  sustainable development
- 可淬硬铸铁凸轮轴
  quenchable cast iron camshaft
- 可见光催化
  visible light photocatalysis
- 可控合成
  controlled synthesis
- 可控气氛
  controlled atmosphere
- 可控氧流冶金
  controlled oxygen flow metallurgy
- 可选性
  washability
- 空白焙烧
  blank roasting
- 空气变形喷嘴
  ATY nozzle
- 空位掺杂
  vacancy doping

- 空心结构

  hollow structure
- 空心球

  hollow spheres
- 空心球 TiO_2

  TiO_2 hollow spheres
- 孔隙率

  porosity
- 孔隙特性

  pore characteristics
- 孔型轧制

  pass rolling
- 控冷

  controlled cooling
- 控冷终止温度

  termination temperature after controlled cooling
- 控轧控冷工艺

  controlled rolling and cooling technology
- 控制轧制

  controlled rolling process
- 库仑滴定

  coulomb titration
- 库仑效率

  coulombic efficiency
- 快速沉积

  rapid deposition
- 快速成形

  rapid prototyping

Kk

- 快速加热回火
  rapid heating tempering
- 宽频带
  broadband
- 宽温
  broad-temperature-range
- 矿化剂
  mineralization
- 矿化远景
  mineralization prospect
- 矿山地质
  mining geology
- 矿山开发
  mine development
- 矿山遗迹
  mining heritage
- 矿石
  ore
- 矿石分解
  ore decomposition
- 矿石特征
  mineral characteristics
- 矿物材料
  mineral materials
- 矿物分解
  mineral decomposition
- 矿物构造
  mineral structure

- 矿物嵌布特性
  mineral embedded properties
- 矿物特性
  mineral property
- 矿物学特征
  mineralogical characteristics
- 矿物组成
  mineral composition
- 矿相
  mineral phase
- 矿相分析
  mineralogy analysis
- 矿相显微
  mineral phase microscopy
- 矿业生态工业园
  ecological mining industrial garden
- 矿渣
  slag
- 框架钛
  framework titanium structure
- 扩大试验
  scale-up test
- 扩散
  diffusion
- 扩散复合
  diffusion bonding
- 扩散激活能
  activation energy of diffusion

Kk

- 扩散连接
  diffusion bonding
- 扩散系数
  diffusion coefficient

L

- 拉拔
  drawing
- 拉曼光谱
  Raman spectra
- 拉曼散射
  Raman Scattering
- 拉伸
  tensile
- 拉伸力学性能
  tensile mechanical properties
- 拉伸试验
  tensile test
- 拉伸性能
  tensile properties
- 拉深成形
  deep drawing
- 拉应力
  axial-stretching stress
- 镧
  lanthanum
- 类钙钛矿型复合氧化物
  perovskite-like oxides
- 类骨磷灰石
  carbonated hydroxyapatite

- 冷等静压
  cold isostatic pressing
- 冷固球团
  cold-bound pellets
- 冷加工
  cold processing
- 冷凝外貌
  condensation appearance
- 冷却方式
  cooling mode of sinter
- 冷却速度
  cooling speed
- 冷却速率
  cooling rate
- 冷速
  cooling velocity
- 冷轧板
  cold rolled sheet
- 离散元分析
  discrete element analysis
- 离子氮化
  plasma nitriding
- 离子电迁移率
  ionic mobility
- 离子交换法
  ion exchange
- 离子交换容量
  ion exchange capacity
- 离子交换色谱
  ion exchange chromatography

钒钛研究关键词中英文手册

- 离子交换树脂
  ion exchange resin
- 离子注入
  ion implantation
- 理化性能
  physical and chemical property
- 理论系数
  therretical coefficient
- 理论燃烧温度
  theoretical combustion temperature
- 锂二次电池
  lithium secondary batteries
- 锂钒氧化物
  lithium vanadium oxide
- 锂离子电池
  lithium-ion batteries
- 力学性能
  mechanical properties
- 利用系数
  utilization coefficient
- 隶属函数
  membership function
- 粒度
  particle size
- 粒度分布
  particle size distribution
- 粒度分析
  granularity analysis
- 粒度特性
  particle size characteristics

- 粒径分布
  particle size distribution
- 粒径可控
  size-controllable
- 粒铁
  iron nuggest
- 连续金相法
  metallographic method of continuous beating
- 连续通道
  continuous-channel
- 连续退火
  continuous annealing
- 连铸
  continuous casting
- 连铸坯
  casting billet
- 帘线钢
  tire cord steel
- 联合法
  joint method
- 炼钢钒渣
  vanadium slag from steel making
- 炼铁
  ironmaking
- 炼铁成本
  ironmaking cost
- 两步法
  two-step approach
- 两步溶胶-凝胶法
  two-step sol-gel method

- 亮绿 SF
  light green SF
- 量子地球化学
  quantum geochemistry
- 量子化学
  quantum chemistry
- 料层
  charge layer
- 裂解
  pyrolysis
- 裂纹
  crack
- 临界区
  critical zone
- 临界应变
  critical strain
- 磷钼钒季铵盐
  Molybdovanadophosphoric quaternary ammonium
- 磷钼酸
  Molybdophosphoric acid
- 磷酸钙
  calcium phosphate
- 磷酸钠
  sodium phosphate
- 磷铁
  ferro phosphorus
- 灵敏度
  sensitivity
- 流变应力
  flow stress

- 流程
  leaching
- 流动海水
  flowing seawater
- 流动性
  fluidity
- 流化床
  circulating fluidized bed
- 流化质量
  fluidization quality
- 流态化
  fluidization
- 硫
  sulfur
- 硫酸法
  sulphuric acid method
- 硫酸法钛白
  sulfate titanium dioxide
- 硫酸化焙烧
  sulphating roasting
- 硫酸浸出
  sulfuric acid leaching
- 硫酸钛
  titanium sulfate
- 硫酸亚铁
  ferrous sulfate
- 硫酸氧钒
  vanadyl sulfate
- 硫酸氧钛
  titanyl sulfate

钒钛研究关键词中英文手册

- 六钛酸钾晶须
  potassium hexatitanate whiskers
- 炉衬
  lining refractory
- 炉底炉缸
  furnace bottom and hearth
- 炉底上涨
  furnace bottom-rising
- 炉缸
  hearth
- 炉缸炉底侵蚀
  hearth erosion
- 炉缸侵蚀
  erosion of hearth
- 炉缸钛沉积量
  buildup amount of titanium at furnace hearth
- 炉料
  furnace burden
- 炉料结构
  burden structure
- 炉龄
  furnace work-life
- 炉体结构
  structure of furnace
- 炉外法
  outside furnace process
- 炉渣碱度
  slag basicity
- 炉渣起泡
  foaming slag

Ll

- 铝
  aluminum
- 铝合金
  aluminum alloy
- 铝热法
  thermite method
- 铝热自蔓延
  thermit reduction-SHS
- 铝酸钠溶液
  sodium aluminate solution
- 铝系钒铁炉渣
  thermite method ferrovanadium slag
- 氯苯
  chlorobenzene
- 氯化
  chloride
- 氯化铵
  ammonium chloride
- 氯化法
  chlorination process
- 氯化法钛白
  chloride TiO_2 pigment
- 氯化废弃物
  chloride waste
- 氯化钠
  sodium chloride
- 氯化钠焙烧
  roasting with salt
- 孪晶
  deformation twin

- 罗丹明
  rhodamine
- 螺旋选矿
  spiral separation
- 络合
  complexing
- 落地贮存
  storage
- 落下强度
  drop strength

M

- 马氏体钢
  martensitic steel
- 马氏体时效不锈钢
  maraging stainless steel
- 脉冲电沉积
  pulsed electrochemical deposition
- 脉冲激光沉积
  pulsed laser deposition
- 毛细管-熔池法
  capillary-molten pool method
- 煤粉
  coal powder
- 镁
  magnesium
- 镁钙锌合金
  Mg-Ca-Zn alloy
- 镁合金
  magnesium alloy
- 镁粒
  magnesium pellet
- 蒙脱石
  montmorillonite
- 锰
  manganese

钒钛研究关键词中英文手册

- 锰盐
  manganese salt
- 锰氧化物
  manganese oxides
- 密度泛函理论
  density functional theory
- 模糊数学
  Fuzzy mathematics
- 模具钢
  die steel
- 模拟江水
  synthetic river water
- 模拟体液
  simulated body fluid
- 膜分离
  membrane separation
- 膜厚
  film thickness
- 摩擦磨损
  friction and wear
- 摩擦磨损性能
  tribological property
- 摩擦性能
  tribological property
- 摩擦学机理
  tribological mechanism
- 磨擦系数
  friction coefficient
- 磨矿分级
  grinding and classification

Mm

- 磨矿细度
  grinding fineness
- 磨球
  grinding balls
- 磨损
  wear
- 磨损机理
  wear mechanism
- 莫来石
  mullite
- 钼
  molybdenum
- 钼、氮元素
  Mo/N element
- 钼钒铅矿
  eosite
- 钼镍钒多金属矿石
  Mo-Ni-V polymetallic ore
- 钼酸铵
  ammonium molybdate

N

- 内固定
  internal fixations
- 内燃机
  internal combustion engine
- 纳滤
  nanofiltration
- 纳米
  nanometer
- 纳米点阵
  nanoarray
- 纳米二氧化钒薄膜
  nano-vanadium dioxide thin film
- 纳米二氧化钛
  titanium dioxide nanoparticle
- 纳米复合材料
  nano-composites
- 纳米钙钛矿
  nano-perovskite
- 纳米管
  nanotubes
- 纳米管径
  nanotube diameter
- 纳米管钛酸
  titanic acid nanotubes

Nn

- 纳米级析出相
  nanometer-sized precipitate
- 纳米结构
  nanostructure
- 纳米金
  nano gold
- 纳米晶
  nano-crystalline
- 纳米晶体
  nanocrystal
- 纳米晶体钛
  nano-grained Ti
- 纳米颗粒
  nanoparticle
- 纳米羟基磷灰石
  nano-hydroxyapatite
- 纳米碳管
  carbon nanotubes(CNTs)
- 纳米碳化钒
  vanadium carbide nanopowders
- 纳米线
  nanowires
- 纳米氧化钛
  TiO_2 nanoparticles
- 耐腐蚀性
  corrosion resistance
- 耐腐蚀性能
  corrosion resistance

- 耐候钢

  weathering steel
- 耐碱液腐蚀

  alkali corrosion resistance
- 耐磨钢

  wear resistance steel
- 耐磨试验

  friction test
- 耐磨性

  wear resistance
- 耐磨硬质相

  wear-resistant hard phase
- 耐蚀性

  corrosion resistance
- 耐酸性

  acid-proof
- 耐酸阳极

  acid-proof anode
- 能带结构

  band structure
- 能量转移

  energy transfers
- 铌钒复合微合金化

  Nb-V micro-alloying
- 铌酸锂

  lithium niobate
- 铌钛微合金化钢

  Nb-Ti microalloying steel

Nn

- 铌铁酸锶-钛酸铅
  strontium iron niobate-lead titanate
- 拟合
  regression
- 逆流浸出
  conunter current leaching
- 逆变换
  reverse transformation
- 黏度
  viscosity
- 镍
  nickel
- 镍钒浸出液
  nickel-vanadium leaching solution
- 镍铬合金
  nickel-chromium alloy
- 镍铬合金
  Ni-Cr alloy
- 镍基耐蚀合金
  nickel-based corrosion resistant alloy
- 镍氢电池
  Ni-MH batteries
- 镍钛
  Ni-Ti
- 镍钛合金
  nickel-titanium alloy
- 镍钛器械
  nickel-titanium instruments

镍钛研究关键词中英文手册

- 镍钛旋转器械
  nickel-titanium rotary files
- 镍钛预备器械
  nickel-titanium endodontic files
- 柠檬酸盐
  citrate
- 凝固
  solidification

O

- 偶联机理
  coupling mechanism
- 偶联剂
  coupling agent
- 耦合
  coupling

P

- 排碱
  alkali discharge
- 攀钢
  Panzhihua Mining Corp.
- 攀西地区
  Panxi region
- 攀西钒钛磁铁
  V-Ti magnetite in Panxi
- 攀枝花低品位钛精矿
  Panzhihua low-grade titanium concentrate
- 攀枝花钒钛磁铁矿矿区
  Panzhihua V-Ti Magnetite Mine
- 泡沫金属
  foamed metal
- 泡沫钛
  titanium foam
- 配合物
  complex
- 配矿
  ore blend
- 配矿比
  proportion
- 配碳量
  carbon adding amount

Pp

- 喷吹
  jetting
- 喷煤
  coal injection
- 喷射鼓泡塔
  jet bubbling reactor
- 喷雾浓缩
  spray concentration
- 喷溅渣
  ejecting slag
- 喷抛丸联合清理
  cimbined shot blasting cleaning
- 喷砂清理
  sand blasting
- 喷丸清理
  shot blasting cleaning
- 喷雾干燥
  spray drying
- 喷雾冷却
  fog hardening
- 喷液淬火
  spray hardening
- 硼化物
  boride
- 硼酸
  boric acid
- 硼钛化合物
  boron-titanium compounds
- 硼铁精矿
  ferroboron concentrate

钒钛研究关键词中英文手册

- 膨润土
  bentonite
- 铍
  beryllium
- 片层间距
  interlamellar spacing
- 片钒
  vanadium pentoxide sheet
- 偏钒酸铵
  ammonium metavanadate
- 偏光棱镜
  polarizing prism
- 偏钛酸
  metatitanic acid
- 偏振转换
  polarization conversion
- 贫锰矿
  less manganese ore
- 品位
  grade
- 平均粒径
  kinetic
- 平均序参数
  average order parameter
- 平面波超软赝势法
  photocatalysis
- 评价
  evaluation
- 评价指标
  evaluation index

Pp

- 破磨参数测定
  crushing-grinding parameters
- 破碎产品磨选性能
  capability of grinding and separating of crushing prodution
- 普通矿
  ordinary mine
- 普通矿冶炼
  ordinary ore smelting
- 普通球团
  ordinary pellet

Q

- 起泡现象
  foaming phenomenon
- 气缸套
  cylinder liner
- 气孔结构
  pore structure
- 气敏性能
  gas sensing property
- 气相氧化
  vapor-phase oxidation
- 钎焊
  brazing
- 铅基复合钙钛矿
  Pb-base complex perovskite structure
- 强磁
  high-intensity
- 强磁选
  high intensity magnetic separation
- 强磁选机
  high intensity magnetic separator
- 强磁预选
  preliminary processing by high-intensity separation
- 强度
  strength

Qq

- 强度指标
  strength index
- 强化措施
  intensification measure
- 强化机理
  strengthening mechanism
- 强化技术
  enhanced technology
- 强化浸出
  strengthening leaching
- 强化冷却
  enhancing cooling effect
- 强化冶炼
  intensified smelting
- 强脉冲离子束
  intense pulsed ion beam
- 强韧化
  strengthening and toughening
- 羟基磷灰石
  hydroxyapatite
- 切削性能
  cutting performance
- 侵蚀
  corrosion
- 钛矿物
  titanium
- 钛铁连侧
  successive deter
- 钛阳极
  titanium anode

钒钛研究关键词中英文手册

- 氢氧化钾
  potassium hydroxide (KOH)
- 清洁生产
  cleaner production
- 球扁钢
  bulb-flat steel
- 球化
  spheroidization
- 球径
  ball diameter
- 球磨改性
  modification by ball-milling
- 球磨活化
  milling activation
- 球墨铸铁
  nodular cast iron
- 球团
  pellet
- 球团矿质量
  pellet quality
- 球形粉末
  spherical powder
- 屈服强度
  yield strength
- 屈氏体磨球
  troostite grinding ball
- 取代型钒氧簇合物
  substituted vanadium-oxide clusters
- 取样
  sampling

- 全钒钛球团
  V/Ti-bearing magnetite pellet
- 全钒钛铁矿冶炼
  smelting with full V-Ti iron ore
- 全钒钛冶炼
  V-Ti-bearing magnetite smelting
- 全钒氧化还原液流电池
  all-vanadium redox flow battery
- 全钒液流电池
  vanadium redox flow battery

R

- 燃耗
  fuel consumption
- 燃料电池
  fuel cells
- 燃料油
  fuel oil
- 热变形
  hot deformation
- 热变形奥氏体
  heat deformed Austenite
- 热冲压成形用钢
  heat stamp-formed steel
- 热处理
  heat treatment
- 热等静压
  hot isostatic pressing
- 热电材料
  thermoelectric materials
- 热返矿率
  hot returns rate
- 热分解法
  thermal decomposition
- 热分析
  thermoanalysis

- 热粉化性
  hot pulverization
- 热风烧结
  hot air sintering
- 热还原
  thermal reduction
- 热极凝固
  heat probe solidfication
- 热浸镀
  hot-dip galvanizing
- 热侵蚀
  hot etching
- 热扩散系数
  thermal diffusivity
- 热偶腐蚀
  thermogalvanic corrosion
- 热喷涂
  thermal spraying
- 热膨胀
  thermal expansion
- 热容
  heat capacity
- 热双金属
  thermostatic biomaterial
- 热稳定性
  thermal stability
- 热压
  hot pressing
- 热浸镀铝
  hot-dip aluminum

钒钛研究关键词中英文手册

- 热浸镀锌
  hot-dip galvanizing
- 热力学
  thermodynamic
- 热力学计算
  thermodynamic calculation
- 热流强度
  thermal flow intensity
- 热膨胀
  thermal expansion
- 热疲劳
  thermal fatigue
- 热水解
  thermal hydrolysis
- 热丝 CVD
  hot filament chemical vapor deposition
- 热稳定性
  thermal stability
- 热效应
  thermal effect
- 热形变
  hot deformation
- 热性能
  characteristic temperature
- 热压缩变形
  hot compression deformation
- 热压氧化
  heating oxidizing
- 热氧化
  thermal oxidation

- 热应力
  thermal stress
- 热影响区
  heat-affected zone
- 热轧带肋钢筋
  hot rolled ribbed reinforced bar
- 热轧复合
  hot-rolling bonding
- 热轧工艺
  hot rolling process
- 热致变色
  thermochromic
- 热重-差热分析
  thermogravimetry and differential thermal analysis
- 热重分析
  thermal gravimetric analysis
- 人工神经网络
  artificial neural network
- 人工唾液
  artificial saliva
- 人牙周膜细胞
  human periodontal ligament cells
- 溶出
  leaching
- 溶剂热
  solvothermal synthesis
- 溶胶凝胶
  sol-gel
- 溶解
  dissolution

- 溶解度
  solubility
- 熔滴试验
  droplet test
- 熔滴性能
  droplet properties
- 熔分
  smelting and separating
- 熔分还原
  melting and reduction
- 熔分渣
  molten slag
- 熔化连接
  fusion bonding
- 熔化区间
  melting temperature range
- 熔化温度
  melting temperature
- 熔化性
  melting property
- 熔化性温度
  melting temperature
- 熔化性质
  melting property
- 熔剂二次分加
  secondary divided flux addition
- 熔剂分加
  flux addition
- 熔坑
  crater

- 熔融
  melting
- 熔融玻璃片
  fusion preparation
- 熔融法
  fusion Method
- 熔融还原
  smelting reduction
- 熔盐
  molten salts
- 熔盐电解
  molten salt electrolysis
- 熔盐氯化
  molten salt chlorination
- 蠕墨铸铁
  vermicular cast iron
- 软化行为
  softening behavior
- 软熔
  cohesive
- 软熔滴落
  softening-melting and dropping
- 软熔滴落带
  cohesive drip belt
- 软熔滴落性能
  softening dropping property
- 软熔性能
  softening-melting properties
- 锐钛二氧化钛
  anatase TiO_2

钒钛研究关键词中英文手册

- 锐钛矿晶型
  anatase crystalline structure
- 润湿性
  wettability
- 弱磁选
  low intensity magnetic separation

S

- 三点弯曲法
  three-point flexure bond test
- 三甲基氯硅烷
  trimethylchlorosilane
- 三维连通孔隙
  3D interconnected pore
- 三维模拟
  3D finite element simulation
- 三维有限元模拟
  3D finite element simulation
- 三维有序大孔材料
  three-dimensionally ordered macroporous material
- 三元无机包膜
  ternay inorganic coated
- 散射截面
  scattering cross section
- 扫描电镜
  SEM
- 烧结
  sintering
- 烧结基础特性
  basic sintering characteristics
- 烧结矿产
  quality of core sintering

钒钛研究关键词中英文手册

- 烧结矿强度
  sinter strength
- 烧结矿性能
  sinter performance
- 烧结矿质量
  sinter quality
- 烧结实践
  practice of sintering
- 烧结试验
  sintering experiment
- 烧结特性
  sintering characteristic
- 烧结条件
  sintering condition
- 烧结脱硫
  sintering desulfurization
- 烧结性能研究
  research sintering
- 烧结制度
  sintering schedule
- 设计特点
  design features
- 射频等离子体
  radio frequency induction plasma
- 深度开发
  deep development
- 深还原渣
  deep reduction slag
- 渗层
  coating

Ss

- 渗钒
  vanadinizing
- 渗硼
  boriding
- 渗透
  infiltration
- 生产
  production
- 生产成本
  production cost
- 生产实践
  operational practice
- 生产线
  production line
- 生长活化能
  growth activation energy
- 生长机理
  growth mechanism
- 生活用水
  domestic water
- 生石灰
  quick lime
- 生态危害
  ecological hazard
- 生物材料
  biomaterials
- 生物活性
  biological activity
- 生物力学相容性
  biomechanical compatibility

钒钛研究关键词中英文手册

- 生物模板法
  biotemplate method
- 生物相容性
  biocompatibility
- 剩磁团聚重选
  residual magnetic agglommeration gravity separation
- 失活机理
  deactivation mechanism
- 湿度
  humidity
- 湿法
  wet method
- 湿敏特性
  humidity sensing property
- 石灰中和
  lime neutralization
- 石煤
  stone coal
- 石煤钠化
  sodiumizing-oxidizing roasting
- 石煤烧渣
  stone coal cinder
- 石煤提钒
  vanadium extraction from carbonaceous shale
- 石油沥青
  asphalt
- 时间序列模型
  time series model
- 时效性能
  aging performance

Ss

- 实践应用
  practical application
- 试验
  experimental
- 适宜值
  reasonable value
- 铈
  cerium
- 铈掺杂
  Ce-doped
- 收缩核模型
  shrinking core model
- 竖炉
  shaft furnace
- 数控弯曲
  NC bending
- 数理统计
  mathematical statistics
- 数值模拟
  numerical simulation
- 双层钙钛矿锰氧化物
  double-layered perovskite manganites
- 双层辉光等离子渗金属技术
  double glow plasma alloying technology
- 双层永磁辊式磁选机
  double-layered roll type magnetic separators
- 双钙钛矿
  double perovskite

钒钛研究关键词中英文手册

- 双钙钛矿结构
  double-perovskite structure
- 双辉等离子低温渗铬
  double glow plasma low temperature chroming
- 双金属复合
  bimetal compound
- 双离子束溅射
  dual ion beam sputtering
- 双相钢
  dual-phase steel
- 双渣法
  double-slag process
- 双锥螺旋分级机
  double-cone spiral classifer
- 水玻璃
  water glass
- 水淬渣
  water quenching slag
- 水解
  hydrolysis
- 水口结瘤
  nozzle clogging
- 水切割
  water jet cutting
- 水热法
  hydrothermal method
- 水热合成
  hydrothermal synthsis

- 水热转化

  hydrothermal transformation
- 水溶液

  aqueous solution
- 水雾化粉末

  water atomized powder
- 水源

  water source
- 四氯化硅

  silicon tetrachloride
- 四氯化钛

  titanium tetrachloride
- 四异丙醇钛

  titanium tetra-isopropoxide
- 松装密度

  apparent density
- 苏长岩

  norite
- 塑模比

  ratio of remains deformation to module
- 塑性变形

  plastic deformation
- 塑性成形

  plastic forming
- 塑性聚合物法

  viscous polymer processing
- 酸催化剂

  acid catalysts

钒钛研究关键词中英文手册

- 酸法浸出
  acid leaching
- 酸回收
  acid recovery
- 酸解
  hydrolysis-acidification
- 酸解率
  acidolysis ratio
- 酸浸
  acid leaching
- 酸浸工艺
  acidic leaching technique
- 酸浸提钒
  extracting vanadium by acid leaching
- 酸浸条件
  leaching process conditions
- 酸浸效率
  leaching efficiency
- 酸浸液
  leaching solution
- 酸浓度
  acid concentration
- 酸溶法
  acid dissolving method
- 酸溶性
  acid-soluble performance
- 酸洗
  pickling

- 酸性铵盐沉淀
  acidic ammonium salt precipitation of vanadium
- 酸性铵盐沉钒
  acidic precipitation of vanadate-leaching solution
- 隧道窑
  tunnelkiln
- 损伤
  damage

钒钛研究关键词中英文手册

T

- 铊
  thallium
- 塔菲尔曲线
  Tafel curve
- 太赫兹波
  terahertz
- 钛
  titanium
- 钛/瓷结合
  titanium-porcelain bonding
- 钛/钢复合板
  titanium clad plate
- 钛/不锈钢复合板
  titanium/stainless steel clad plate
- 钛/钢双金属
  titanium/steel bimetal
- 钛白
  titanium pigment
- 钛白废酸
  titanium white waste acid
- 钛白粉
  titanium pigment
- 钛板
  titanium plate

- 钛宝石
  Ti sapphire
- 钛宝石激光器
  Ti sapphire laser
- 钛材
  titanium
- 钛瓷
  titanium porcelain
- 钛带
  titanium strip
- 钛镀层
  titanium deposit
- 钛钢真空-轧制复合板
  Ti/steel vacuum-rolling clad plate
- 钛锆合金
  Ti-Zr alloy
- 钛锆膜
  Ti-Zr films
- 钛硅薄膜
  TiO_2-SiO_2 film
- 钛硅复合光催化剂
  TiO_2/SiO_2 composite photocatalyst
- 钛硅树脂
  titanium-contained silicone resin
- 钛过渡层
  buffer layer of titanium
- 钛合金
  titanium alloy
- 钛合金铸件
  casting titanium alloy

钒钛研究关键词中英文手册

- 钛辉石
  titanaugite
- 钛基复合材料
  Ti-based composite
- 钛基复合材料
  titanium matrix composite
- 钛基复合材料涂层
  titanium matrix composite coating
- 钛基金属氧化物阳极
  Ti based metal oxide anode
- 钛基生物材料
  Ti-based biomaterial
- 钛基体
  titanium base
- 钛基柱撑黏土
  Ti-pillared interlayered clays
- 钛及钛合金
  titanium and titanium alloys
- 钛及钛合金板坯
  titanium and titanium alloy slabs
- 钛夹
  titanium clips
- 钛金属
  titanium
- 钛精矿
  titanium concentrate
- 钛颗粒
  titanium particles
- 钛矿
  ilmenite

- 钛缆系统
  titanium cable syste m
- 钛离子注入
  titanium ion implantation
- 钛-铝复合板
  titanium-aluminum clad sheet
- 钛铝金属间化合物
  titanium aluminides spherical powde
- 钛膜
  titanium film
- 钛钼复合析出
  complex carbonitrides of titanium and molybdenum
- 钛纳米粒子
  nano titanium
- 钛铌复合微合金化
  titanium and niobium complex microalloying
- 钛镍合金
  Ti-Ni alloy
- 钛-镍合金层
  Ti-Ni alloyed layer
- 钛片
  titanium plate
- 钛平衡
  titanium balance
- 钛溶胶
  titanium sol
- 钛溶解度
  titanium solubility
- 钛石膏
  titanium gypsum

钒钛研究关键词中英文手册

- 钛酸铋钠
  sodium bismuth titanate(BNT)
- 钛酸钾
  potassium titanate
- 钛酸铝
  alumina titanate
- 钛酸钠
  sodium titanate
- 钛酸锶
  strontium titanate
- 钛酸盐
  titanate
- 钛酸盐纳米管
  titanate nanotube
- 钛酸酯偶联剂
  titanate coupling agent
- 钛提取
  titanium extraction
- 钛铁矿
  titanic iron ore
- 钛铁矿浮选
  flotation of ilmenite
- 钛脱氧
  Ti-deoxidized
- 钛微合金钢
  Ti-microalloyed strip
- 钛尾矿再选
  Re-concentration of ilmenite tailings
- 钛系嵌入化合物
  titanic insertion compound

Tt

- 钛阳极
  titanium anode
- 钛液
  titaniferous solution
- 钛液净化
  titanium liquid purification
- 钛源
  titanium source
- 钛渣
  titanium slag
- 钛值
  titanium value
- 钛种植体
  titanium implants
- 钛柱撑膨润土
  Ti-pillared montmorillonite
- 钛铸件
  titanium castings
- 钛资源
  vanadium resources
- 炭气凝胶
  carbon aerogel
- 炭质岩页型钒矿石
  carbonaceous shale navajoite ore
- 碳氮化钒
  vanadium carbides and nitrides
- 碳氮化钛
  titanium carbonitride
- 碳氮化物
  carbonitride

钒钛研究关键词中英文手册

- 碳钢
  carbon steel
- 碳化钒
  vanadium carbide
- 碳化工艺
  carbonization process
- 碳化钛
  titanium carbide
- 碳化物
  carbide
- 碳化渣
  carbide slag
- 碳还原
  carbon reduction
- 碳热还原
  carbothermal reduction
- 碳酸钙
  calcium carbonate
- 碳酸钠焙烧
  sodium carbonate-roasting
- 碳纸
  carbon paper
- 陶瓷
  porcelain
- 陶瓷产率
  ceramic yield
- 陶瓷涂层
  ceramic coating
- 陶瓷原料
  ceramic materials

- 特定元素
  special elements
- 特征温度
  characteristic temperature
- 梯度变化
  change of gradient
- 提纯度
  purity
- 提钒
  extraction of vanadium
- 提钒工艺
  vanadium extraction techniques
- 提钒尾渣
  extracted vanadium tailings
- 提高铁份
  improving iron content
- 提取
  extraction
- 提钛
  extraction of titanium
- 提铁降硅
  iron raise and silicon reduction
- 体瓷
  dentin porcelain
- 体积变化率
  rate of volume change
- 体积电阻率
  volume resistivity
- 体胀
  swelling

钒钛研究关键词中英文手册

- 替位氮
  substitutional nitrogen
- 天然钒钛磁铁矿
  natural vanadium-titanium magnetite
- 天然沸石
  natural zeolite
- 添加剂
  additive
- 铁(钒)精矿
  reducing arsenic
- 铁
  iron
- 铁电薄膜
  ferroelectric thin films
- 铁电体
  ferroelectrics
- 铁电性能
  ferroelectricity
- 铁粉
  iron powder
- 铁晶粒
  metal iron crystal
- 铁矿指标
  iron index
- 铁品位分布
  distribution of iron grade
- 铁水
  hot metal
- 铁水脱硫
  hot metal desulphurization

- 铁水质量

  hot metal quality
- 铁素体

  ferrite
- 铁素体不锈钢

  ferritic stainless steel
- 铁酸钙

  calcium ferrite
- 铁酸盐相

  ferrite phase
- 铁损

  loss of hot metal in slag
- 铁损形态

  iron loss form
- 铁钛磷化物

  ferrotitanium phosphides
- 铁钛平行分选

  parallely separating Fe and Ti
- 铁型

  iron mold
- 铁浴

  iron bath
- 铜

  copper
- 铜冷却壁

  copper cooling stave
- 铜中间层

  copper interlayer
- 透过率

  transmittance

钒钛研究关键词中英文手册

- 透明导电薄膜
  transparent conducting films
- 透气性
  permeability
- 透射比
  transmissivity
- 涂层
  coating
- 涂层厚度
  coatings thickness
- 涂层结构
  performance of coating
- 土木工程
  civil engineering
- 土壤
  soil
- 团球化
  spheroidizing
- 退火
  annealing
- 退火温度
  annealing temperature
- 脱硅
  silicon removal
- 脱硫
  desulphurization
- 脱硫机理
  desulfurization mechanism
- 脱硫率
  desulphurizing rate

Tt

- 脱硫效率
  desulphurization efficiency
- 脱硫渣
  desulfurized slag sinter
- 脱铝
  dealumination
- 脱氢作用
  dehydrogenation
- 脱钛
  titanium removal
- 脱碳
  decarburization
- 脱碳层
  decarburized zone
- 脱位
  dislocations

- 弯曲
  bending
- 弯曲应力
  bending stress
- 弯折带
  kink band
- 危害指数
  risk index
- 微波
  microwave
- 微波辐射
  microwave radiation
- 微波还原
  microwave reduction
- 微波加热
  microwave heating
- 微波介质陶瓷体系
  microwave dielectric ceramics system
- 微波消解
  microwave digestion
- 微测辐射热计
  microbolometer
- 微动磨损
  fretting wear

Ww

- 微观机理
  micromechanism
- 微合金化刚
  microalloyed steel
- 微合金化
  microalloying
- 微合金化钢
  microalloyed steel
- 微合金化重轨钢
  microalloying heavy rail steel
- 微合金铁粉
  microalloy iron powser
- 微合金析出相
  microallying carbonitride precipitation phase
- 微合金元素
  microalloying element
- 微弧氧化
  micro-arc oxidation
- 微弧氧化
  micro-arc oxidation
- 微量
  trace
- 微量钒
  trace vanadium
- 微量元素
  trace elements
- 微量元素钒钛
  trace element vanadium-titanium
- 微量元素共掺杂
  trace element-codoping

143

钒钛研究关键词中英文手册

- 微裂纹
  micro-crack
- 微孪晶栅
  micro-twin gate
- 微米球
  mciro-sphere
- 微球
  spherical micro particles
- 微乳液
  microemulsion
- 微生物燃料电池
  microbial fuel cell
- 微钛熔渣
  slags with minimal titanium dioxide
- 微型钢（钛）板
  ministeel (titanium) plate
- 微型钛板
  titanium microplate
- 伪装隐身
  camouflage and stealth
- 尾渣
  tailings
- 温度
  temperature
- 温度补偿
  temperature compensation
- 温度场
  temperature field
- 温度效应
  temperature effect

Ww

- 温降
  temperature drop
- 稳定性
  stability
- 稳频
  frequency stabilization
- 温压
  warm pressing
- 稳定化处理
  stabilizing treatment
- 稳定化退火
  stabilizing annealing
- 污染层
  surface contamination layer
- 污染防治
  pollution control
- 污染物负荷指数
  pollution loading index(PLI)
- 钨掺杂
  tungsten doped
- 无缝管
  seamless tube
- 无铬
  chromium-free
- 无机钛硅源
  inorganic titanium and silicon source
- 无筛板沸腾氯化
  fluidized-bed chlorination without sieve plate
- 无限冷硬铸铁
  infinitely chilled cast iron

钒钛研究关键词中英文手册

- 五氧化二钒
  vanadium pentoxide
- 物理光学
  physical optics
- 物理化学性能
  physical and chemical properties
- 物料平衡
  material balance
- 物相
  phase
- 物相测定
  phase measurement
- 物相分析
  phase analysis
- 物相组成
  phase composition
- 物性分析
  physical property analysis
- 物质组成
  material composition
- 误差分析
  error analysis

X

- 吸附模型
  adsorption model
- 吸附容量
  adsorption capacity
- 吸附行为
  adsorption behavior
- 吸附铀
  adsorption uranium
- 吸光光度法
  absorption spectrophotometry
- 吸收光谱
  absorption spectrum
- 吸收截面
  absorption cross section
- 希夫碱
  Schiff base
- 昔格达地层
  Xigeda Strata
- 析出强化
  precipitation strengthening
- 析出物
  precipitates
- 析出相
  precipitated phase

钒钛研究关键词中英文手册

- 析金性能
crystallization property
- 稀土
rare earths
- 稀土发光材料
rare-earth luminescence materials
- 稀土钒共渗
rare earth vanadizing
- 稀土氧化物
rare earth oxide
- 细胞膜
cell membrane
- 细观本构模型
meso-scale constitutive model
- 细晶强化
refined crystalline strengthening
- 细磨
fine grinding
- 显微硬度
micro-hardness
- 显微组织
microstructure
- 现状及进展
status and progress
- 线粒体
mitochondria
- 线膨胀率
linear expansion rate
- 线性化
linearization

Xx

- 相变
  doping
- 相变压力
  phase transition pressure
- 相成分
  phase composition
- 相关关系
  correlation
- 相关性
  correlation
- 相关性分析
  correlation analysis
- 相间沉淀
  interphase precipitation
- 相结构
  phase component
- 消费
  consumption
- 消解方法
  digestion method
- 小波变换法
  spectrophotometry
- 偕胺肟化反应
  amidoxime reaction
- 斜长石
  plagioclase
- 斜长岩
  plagioclase
- 锌-镍-钒合金
  zinc-nickel-vanadium alloy

钒钛研究关键词中英文手册

- 新的操作方法
  new operation process
- 新工艺
  new process
- 新西兰海砂
  New Zealand sea sand
- 新西兰矿粉
  New Zealand fine ore
- 形成能
  formation energy
- 形核孕育期
  nucleation incubation period
- 形貌表征
  research progress
- 形貌理论预测
  theoretic predictions of morphology
- 形貌演化
  morphological evolution
- 形态
  speciation
- 形态分布
  species distribution
- 形态转化
  speciation transformation
- 形状记忆效应
  shape memory effect
- 形状记忆性能
  shape-memory behavior
- 性能
  property

Xx

- 性能检测与评价
  performance test and evaluation
- 性能碳当量
  carbon equivalent of property
- 锈层
  rust layer
- 锈蚀速率
  rust rate
- 絮凝
  flocculation
- 絮凝剂
  flocculant
- 悬浮物
  suspended substance
- 旋流电解
  rotational flow electrodeposition
- 选矿成本
  beneficiation cost
- 选矿方法
  mineral processing method
- 选矿流向
  flow direction during beneficiation
- 选螺旋溜槽-摇床重选
  spiral chute-table gravity separation
- 选钛
  ilmenite beneficiation
- 选铁
  iron separation
- 选铁工艺
  iron separation technology

钒钛研究关键词中英文手册

- 选铁工艺流程
  iron separation technological flowsheet
- 选铁试验
  experients of dressing iron
- 选铁尾矿
  tailings after titanomagnetite recovery
- 选择性催化还原
  selective catalytic reduction (SCR)
- 选择性析出
  selective precipitation
- 选择氧化
  selective oxidation
- 循环变形行为
  cyclic deformation
- 循环富集
  enrichment by circulating
- 循环利用
  recycling
- 循环流化床
  circulating fluidized bed
- 循环氧化
  cyclic oxidation

Y

- 压电陶瓷
  piezoelectric ceramics
- 压电性能
  piezoelectric property
- 压球
  pellet pressing
- 压团工艺
  briquetting process
- 压致相变
  pressure-induced phase transition
- 亚结构
  substructure
- 亚熔盐
  sub-molten salt
- 亚熔盐法
  sub-molten salt method
- 亚熔盐反应
  sub-molten salt reaction
- 亚细胞分布
  subcellular distribution
- 烟气
  flue gas
- 烟气脱硫
  flue gas desulfurization

钒钛研究关键词中英文手册

- 岩浆分异
  magmatic differentiation
- 岩浆矿床
  magmatic deposit
- 岩浆热液
  magmatic hydrothermal fluid
- 岩浆晚期贯入
  late magma injection
- 岩浆晚期结晶分异
  late magmatic crystallization differentiation
- 岩相分析
  petrographic analysis
- 岩心钻探
  core drilling
- 炎症因子
  inflammatory cytokines
- 研究进展
  research progress
- 研磨预处理
  abrading pretreatment
- 盐酸浸取
  hydrochloric acid leaching
- 盐雾腐蚀
  salt spray corrosion
- 验证试验
  verification test
- 阳光控制性能
  solar control performance
- 阳极极化
  anodic polarization

Yy

- 阳极氧化
  anodic oxidation
- 阳极氧化膜
  anodic oxide film
- 氧含量
  oxygen content
- 氧化
  oxidation
- 氧化焙烧
  oxidizing roasting
- 氧化处理
  oxidation treatment
- 氧化动力学
  oxidation kinetics
- 氧化钒薄膜
  vanadium oxide films
- 氧化反应
  oxidation
- 氧化钙
  calcium oxide
- 氧化锆
  zirconia
- 氧化机理
  oxidation mechanism
- 氧化铝
  alumina
- 氧化铝空心球
  alumina hollow ball
- 氧化膜
  oxidation film

- 氧化球团
  oxidized pellets
- 氧化球团矿
  oxide pellet
- 氧化时间
  oxidation time
- 氧化钛
  titania
- 氧化钛纳米管层
  titanium oxide nanotubes layer
- 氧化脱氢
  oxidative dehydrogenation
- 氧化物涂层
  oxide coatings
- 氧化物冶金
  oxide metallurgy
- 氧化性能
  oxidative property
- 氧化铱
  iridium oxide
- 氧扩散层
  oxide-rich layer
- 氧压
  oxygen pressure
- 氧逸度
  oxygen fugacity
- 冶金技术
  metallurgical technology
- 冶金结合
  metallurgical bonding

Yy

- 冶金性能
  metallurgical property
- 冶炼
  smelting
- 冶炼强度
  smelting strength
- 液流电池
  flow battery
- 液流钒电池
  vanadium redox flow battery
- 液相烧结
  liquid phase siutering
- 液压胀形工艺
  hydroforming technology
- 乙酸
  acetic acid
- 以废治废
  waste circulation
- 异化还原
  dissimilatory reduction
- 抑制
  depression
- 抑制机理
  inhibit mechanism
- 抑制剂
  inhibitor
- 抑制作用
  inhibition
- 易熔玻璃
  fusible glass

钒钛研究关键词中英文手册

- 阴极
  cathode
- 阴极材料
  cathodic material
- 阴极微弧电沉积
  cathodic microarc electrodeposition
- 银
  silver
- 应变速率
  strain rate sensitivity
- 应力
  stress
- 应用
  application
- 应用前景
  application prospect
- 荧光粉
  luminescent powder
- 荧光性质
  fluorescence property
- 影响
  effect
- 硬度
  hardness
- 硬面合金
  hard facing alloy
- 硬质薄膜
  hard films
- 永磁磁选机
  permanent magnetic separator

Yy

- 优化配矿
  optimization of matching ores
- 优化配料
  burden optimization
- 优先浮选
  selective flotation
- 油酸钠
  sodium oleate
- 有机酸
  organic acid
- 有价残余元素
  valuable residual element
- 有限元
  finite element
- 有限元模式
  finite element analysis
- 诱导铁素体
  ferrite nucleation stimulation
- 预报模型
  forecasting model
- 预成形钛网
  preformed orbital titanium mesh
- 预处理
  pretreatment
- 预处理方法
  pretreatment methods
- 预防方法
  prevention methods
- 预分选
  pre-separation

钒钛研究关键词中英文手册

- 预分选精矿
  pre-separation concentrate
- 预还原
  pre-reduction
- 预还原球团
  pre-reduced pellets
- 预抛尾
  preconcentration
- 预氧化
  pre-oxidation
- 预制粒
  prefabricated grain
- 元素
  elements
- 原胞自动机
  cellular automaton
- 原料
  material
- 原位反应
  in-situ reaction
- 原位合成
  in-situ processing
- 原位生成复合材料
  in-situ synthesized composite
- 原因分析
  reason analysis
- 原子发射光谱分析
  atomic emission spectrometric analysis
- 原子互扩散
  atomic interdiffusion

Yy

- 原子吸收光谱法
  flame atomic absorption spectrometry
- 圆筒制粒机
  mixing drum
- 云母钛
  mica-titania

Z

- 杂多阴离子
  heteropoly anion
- 杂质元素
  impurity elements
- 再加热温度
  reheating temperature
- 再结晶
  recrystallization
- 再结晶激活能
  recrystallization activation energy
- 再结晶温度
  recrystallization temperature
- 再生
  regeneration
- 再生放大
  regenerative amplification
- 造块
  agglomeration
- 造渣制度
  slag forming system
- 造纸废水
  paper making wastewater
- 噪声
  noise

Zz

- 渣皮
  slag skull
- 渣铁分离
  separation of hot metal from slag
- 渣系
  slag system
- 渣相结构
  slag structure
- 轧制复合
  roll bonding
- 轧制温度
  rolling temperature
- 粘度
  viscosity
- 粘结剂
  binder
- 占位填料法
  occupied packing method
- 战略开发
  strategic exploration
- 针状铁素体
  acicular ferrite
- 真空
  vacuum
- 真空还原精炼
  vacuum reduction refining
- 真空蒸馏
  vacuum distillation
- 蒸馏
  distillation

钒钛研究关键词中英文手册

- 蒸汽预热
  preheating by steam
- 整合
  integration
- 正硅酸乙酯
  tetraethyl orthosilicate
- 正极材料
  cathode material
- 正极电对
  positive couple
- 正交试验
  orthogonal experiment
- 正交组合
  orthogonal sintering experiment
- 直接还原
  direct reduction
- 直接还原历程
  direct reduction process
- 直接还原铁
  direct reduction iron
- 直接酸浸
  direct acid leaching
- 直流磁控溅射
  DC magnetron sputtering
- 直流电弧炉
  direct current electric arc furnace
- 直流反应磁控溅射
  reactive DC magnetron sputtering
- 植物有效性
  phyto-availability

Zz

- 纸片法
  disk diffusion method
- 指标
  index
- 酯化
  esterification
- 制备
  preparation
- 制备策略
  synthetic strategies
- 制备工艺
  preparation processing
- 制粒效果
  granulation effect
- 制粒效率
  granulating efficiency
- 制粒性指数
  granulating index
- 治理
  disposal
- 质子电导率
  proton conductivity
- 质子交换膜
  proton exchange membrane
- 中低冲击载荷
  medium-low impact load
- 中铬白口铸铁
  medium chromium white cast iron
- 中硅铸铁
  medium silicon cast iron

钒钛研究关键词中英文手册

- 中和
  neutralization
- 中间层
  inner layer
- 中空微球
  hollow microsphere
- 中矿筛分
  middling screening
- 中锰球墨铸铁
  medium manganese nodular cast iron
- 中温
  moderate temperature
- 中心偏析
  center segregation
- 中性红
  neutral red
- 中子管
  neutron tube
- 种植体
  implants
- 重金属
  heavy metal
- 重金属废水
  wastewater containing heavy metal ions
- 重铬酸钾
  potassium dichromate
- 轴承钢
  bearing steel
- 轴对称偏振光
  cylindrical vector beams

- 珠光体
  pearlite
- 珠光颜料
  pearlescent pigment
- 主成分分析
  principle component analysis
- 主次量组分
  major and minor components
- 主量元素
  major elements
- 贮氢合金
  hydrogen storage alloy
- 柱
  column
- 柱撑膨润土
  pillared bentonite
- 铸钢
  cast steel
- 铸态中锰钢
  as-cast medium manganese steel
- 铸铁
  cast iron
- 转变分数
  mean diameter
- 转底炉
  rotary hearth furnace
- 转鼓强度
  drum strength
- 转化膜
  conversion coating

钒钛研究关键词中英文手册

- 转炉
  converter
- 转炉炼钢
  converter steelmaking
- 转运
  translocation
- 装置
  equipment
- 资源高效利用
  efficient utilization of resource
- 资源利用
  exploitation
- 资源现状
  resources situation
- 资源综合利用
  comprehensive utilization of resources
- 自放电
  self discharge
- 自蔓延高温合成
  self-propagating high-temperature synthesis
- 自磨
  autogenous grinding
- 自相似方法
  self-similarity method
- 综合回收
  comprehensive recovery
- 综合回收钛
  comprehensive recovering titanium
- 综合利用
  comprehensive utilization

Zz

- 综合能耗
  comprehensive energy consumption
- 阻抗特性
  impedance charaeteristics
- 组合浸出剂
  composed leaching reagent
- 组织
  microstructure
- 组织和性能
  microstructure and properties
- 组织及性能
  structure and property
- 组织模拟
  tissue simulation
- 最小流化速度
  minimum fluidization velocity
- 唑系配体
  zole-like ligands